Are We Moving to

By Anne Schraff

Illustrations by Michael Carroll

John Muir Publications
Santa Fe, New Mexico

For Andy and Chris and all the brave young dreamers.

Special thanks to Gregory Vogt, National Aeronautics and Space Administration (NASA); Dr. Richard Taylor, British Interplanetary Society (BIS); and Dr. Yoji Ishikawa.

John Muir Publications, P.O. Box 613, Santa Fe, NM 87504
Text and illustrations copyright © 1996 by John Muir Publications
All rights reserved.

Printed in the United States of America
First edition. First printing October 1996

Library of Congress Cataloging-in-Publication Data
Schraff, Anne E.
    Are we moving to Mars? / by Anne Schraff.
        p.   cm.
    Summary: Presents different proposals that have been advanced about the colonizing of Mars.
    ISBN 1-56261-310-3
    1. Mars (Planet)—Surface—Juvenile literature. 2. Life on other planets—Juvenile literature. 3. Extraterrestrial (anthropology—Juvenile literature. [1. Mars (Planet) 2. Planets—Environmental engineering. 3. Life on other planets.] I. Title.
QB641.S28   1996
620'.419—dc20
      96-22650
          CIP
          AC

Editors: Rob Crisell, Peggy Schaefer
Design: Marie J.T. Vigil
Production: Marie J.T. Vigil, Nikki Rooker
Printing: Guynes Lithographers

Distributed to the book trade by
Publishers Group West
Emeryville, California

Front cover: A scientist explores Mars in his rover. Behind him is a Martian settlement.
Title page: From one of Mars' two moons, astronauts get a good view of the Red Planet.
Right: Mars as seen in 1976 by Viking I

NASA

Are We Moving to Mars?

*What would you say to a Martian,
If a Martian came into view?
Just take a look in the mirror,
That Martian may someday be you!*

When you look into the night sky and see a red glow brighter than any star, you're probably looking at Mars. Mars is more like Earth than any other planet in our solar system, but it's also very different from our planet in many ways. After seeing close-up photos of the planet, some people say it looks like an orange with grayish green spots. We've been looking out at Mars and wondering about it for thousands of years. That might be why we're so eager to plant our feet in that red Martian soil and gaze up at that pink Martian sky. For many years, going to Mars was the dream only of science fiction writers and movie directors. In the future, though, hundreds of thousands of us might actually be moving to Mars! This book examines the opinions of a few scientists who hope their dream of living on Mars becomes a reality.

Looking at this satellite photo, you can see why Mars is often called the "Red Planet." →

What We Used to Think about Mars

For a long time, people thought alien creatures lived on Mars. Even though they didn't look like us, we imagined that Martians were intelligent and far more advanced than humans. Martians in popular movies and novels often had little bodies with spindly arms and legs. Since they were smarter than us, they needed big heads for their big brains. Of course, they didn't need many muscles because they used robots to do all their work. In Hollywood, some movie aliens were nice, like that famous space creature E.T. Others were monsters determined to steal Earth away from us. Although we were good at making up stories about Martians, we really had no idea if any existed. In fact, we didn't know if there was any life at all on the "Red Planet," as Mars is sometimes called.

This creature from an old movie called *This Island Earth* shows how Hollywood used to imagine visitors from outer space. Judging from his looks, this alien isn't too happy to be on Earth!

What We Know Now

People have learned a lot since we first began thinking about Martians. Today, we know that Mars is not like any place in the United States, Europe, or Asia. The temperature on Mars can get down to 195 degrees below zero—ten times colder than Antarctica. Because the planet doesn't have much oxygen, the air is too thin to breathe. Mars is a desert planet with no liquid water and no sign of life. Gravity is low, too. If you walked on Mars, you could take springing steps, as if you were on a mattress. If you can jump three feet high on Earth, you would be able to jump nearly nine feet high on Mars!

So, even though Mars is different from our planet in important ways, is there a way we could still live there?

Mars is our closest neighbor. It's more similar to Earth than any other planet in our solar system. When you compare Mars and Earth to planets like Jupiter and Saturn, they are tiny!

Terraforming: Creating Another Earth

Terraforming means making a planet look and feel like Earth. Terraforming some planets would be impossible. Venus, for example, has thousands of lava-spewing volcanoes and an atmosphere that keeps the planet's surface at about 900 degrees Fahrenheit. Trying to terraform its surface would be almost impossible. Venus is just too different from Earth.

Mars is another story. Even though Mars is very cold and its atmosphere is only 1/100 as dense as Earth's, some scientists think we could make it more breathable by shooting special gases into the Martian air. Giant mirrors in space could warm the surface of Mars by reflecting sunlight onto its surface. We could explode bombs more powerful than any ever used, freeing water and minerals from the Martian soil.

All this work might eventually transform Mars into a warm, green, moist planet like Earth. But all this terraforming would take a very long time, perhaps as long as 10,000 years. Other scientists say there is another way to get humans on Mars within only 100 years. They call their scheme *para-terraforming*.

On this page, you can see what Mars might look like after thousands of years of terraforming.

Para-Terraforming: Speeding up the Process

Para-terraforming means building structures on Mars that are capable of sheltering people against the planet's thin atmosphere and bitter cold. Some para-terraformers have imagined structures that might look like enlarged spaceships. Others have even grander plans, such as building giant roofs over the planet and towering skyscrapers that rise two miles into the sky. This method of transforming Mars would take far less time compared to terraforming, but would it be better?

This book will show you how our ideas of Mars have changed over the centuries, how we learned what Mars is truly like, and how a few scientists hope to get people to Mars in less than 50 years!

This illustration shows what Mars might look like after hundreds of years of para-terraforming.

Mars through the Ages

For many ancient civilizations, Mars was a mysterious red traveler in the night sky, inspiring curiosity and fear. The ancient Romans gave the planet its name, naming it after Mars, their god of war. In Roman myths, Mars is always looking for a fight. He wears a shiny helmet, a leather shield on his arm, and carries a bronze spear. He's Batman, Superman, and the Terminator rolled into one. Like the superheroes of today, he's very handsome, too, and strong enough to defeat all his enemies.

Superstition about Mars continued long after Roman times. During the

This poster from the 1952 movie version of *The War of the Worlds* **shows that people were interested in Martians at this time.**

Mars was the Roman god of war. As you can see from this antique drawing, Mars never goes anywhere without his trusty spear and shield.

1800s, many people viewed the appearance of Mars in the night sky as an omen of war. Astronomers pointed out that Mars was close to Earth at the time of the American Civil War and during the Franco-Prussian war of 1871. As Mars became more frightening and more fascinating, writers began to use it in their stories. We didn't have the technology to visit Mars, but we used our imaginations to take us there.

The writer H.G. Wells wrote a book called *The War of the Worlds* in 1898. In it, he described a deadly invasion of Earth by angry Martians. Wells painted a gruesome picture of the invading Martians:

> "Two large, dark-colored eyes regarded me steadfastly. The mass that framed them . . . was rounded and had, one might say, a face. There was a mouth under the eyes, the lipless brim of which quivered and panted and dropped saliva. The whole creature heaved and pulsated convulsively."

In 1938, the clever actor and director Orson Welles turned *The War of the Worlds* into a radio drama broadcast, producing it as though it were a

In 1938, actor Orson Welles (standing with his arms raised) scared many radio listeners across the U.S. with his dramatic presentation of *The War of the Worlds*. Many listeners thought Martians had actually landed!

real news program. Many listeners thought they were hearing a bulletin about Martians landing in New Jersey. Terrified people fled their houses, jumped into their cars, and tried to escape. The streets were jammed with frightened Americans who believed Earth had been captured by Martians. Was it really that easy to fool so many people? Yes, it was, because lots of folks believed that intelligent creatures lived on Mars.

Scientists Search the Skies

Aristotle, a Greek thinker who lived about 2,300 years ago, gave one of the first scientific opinions about Mars. He noticed that Mars would often disappear behind the moon, so he realized it was farther from Earth than the moon. When Galileo invented the telescope in 1609, astronomers could finally study Mars more closely. Little by little, they made discoveries about the orbit of Mars and even the planet's surface. When English astronomer William Herschel looked at Mars in 1771, he became convinced that the planet was similar enough to Earth to support life. Suddenly, a scientist was talking about living beings on Mars! Herschel had set the stage for wacky new theories about the Red Planet.

The Martian Canal-Builders

In 1877, Italian astronomer Giovanni Schiaparelli observed dark lines on the surface of Mars. He called these lines *canali* or channels. Even though Schiaparelli never said these channels had been built by intelligent beings, soon the whole world was talking about the amazing Martian-built canals on Mars. Before long, people believed Martians were clever engineers, able to build giant canals for their crops and cities. This led to the theories of the scientist some people called "Canal Man"—Percival Lowell.

Percival Lowell was one of the first astronomers to believe intelligent creatures lived on Mars.

American astronomer Percival Lowell was one of the scientists who supported the canal theory. In 1894, he said that if canals existed on Mars, they must have been built by living creatures. Therefore, Martians must exist. In his books about Mars, Lowell described Martians in exact detail. Some of his ideas might seem silly to us now, but Lowell and lots of other scientists didn't think they were. Lowell believed that:

- Martians were far more clever than people on Earth because Mars is older than Earth.
- Martians had inventions that would amaze humans. Besides telephones and movies, they had invented even more wonderful gadgets beyond our wildest imaginations.
- Martians used canals to bring water from the planet's polar caps to the dry regions near Mars' equator. Martian cities were located where canals crossed one another.

This might be what Percival Lowell imagined when he wrote about cities on Mars. Here, the artist has drawn the canals that Lowell believed brought water into Martian cities.

Science Silences the Martian Rumors

Not all scientists accepted Lowell's ideas. In fact, most people made fun of them. Still, the idea that Martians existed sparked the public's imagination. Some people even worried that Martians were suffering from a planet-wide drought. They talked of schemes to communicate with Mars in order to help its inhabitants.

During the 1900s, stronger telescopes allowed astronomers to get better looks at Mars. These new peeks at the red, rocky surface of the planet seemed to show that not only was there no intelligent life on Mars, there probably wasn't any life at all. Schiaparelli and Lowell's canals might have held water at one time, but now they were permanently frozen glaciers.

Still, just because no one lives on Mars doesn't mean someone might not live there in the future. For instance—us!

Getting There: Why, When, How?

Humankind's explorations are limited only by our imaginations. We have crossed oceans, climbed mountains, journeyed to the depths of seas, and learned to fly. We have even walked on the Moon and built space stations in Earth's orbit. But there are many people who wonder if we should even try to go to Mars. Do you think we should? Here are a few reasons for going that you might not have thought of:

- **To Build a Space Station.** If we build a space station on or near Mars, we could use it as a launch pad from which we could visit other planets.

← **Polar ice caps on Mars might contain enough water for humans to survive on the Red Planet.**

Humans have always been daring explorers. In 1969, we landed on the moon. Could Mars be our next adventure?

• **To Learn about Ourselves.** By making Mars warm and green with life, we would learn a lot about keeping Earth livable. We think Mars' atmosphere was almost destroyed billions of years ago, leaving the planet cold and lifeless. Many scientists say that this is exactly what is happening on Earth today. There are holes in the Earth's ozone layer, which is the layer of gases that protects Earth from most of the sun's dangerous radiation. If the holes get too big, our atmosphere won't protect us from the sun anymore. By learning how to improve Mars' atmosphere, we could learn to help our own.

• **To Make a New Home.** Mars could be a new, exciting frontier for the people who want to live there. As Earth gets more crowded, Mars might look like a better place to us. After all, there won't be any traffic jams or overcrowding on Mars!

The Russian *Mir* space station is humankind's first home in space. Building a space station on Mars would be a much greater challenge.

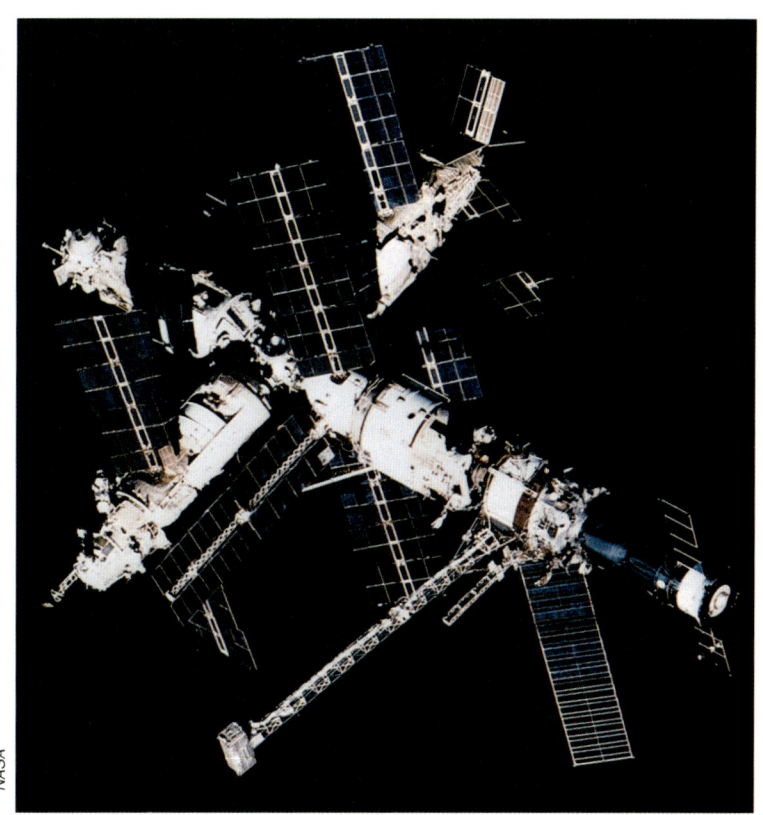

Our First Look at the Red Planet

When we received the first close-up photos of Mars from the United States satellite *Mariner 4* in 1965, the results dismayed lots of us. Why? As *Mariner* sailed a mile above the Red Planet, we were able to see at last what a deserted, rocky place Mars really was. Mars looked more like the Moon than it did Earth! This Mars was nothing like the Mars that science fiction movies had prepared us to see.

A rocket ship launches one of the *Viking* satellites into space in 1976. The *Viking* satellites have brought back most of the photos we have of the Red Planet.

Mars is much more than a flat desert. The *Viking* and *Mariner* satellites have taken pictures of mountains, canyons, craters, and ice caps.

Then, in 1971, *Mariner 9* flew by Mars. It took photos of new areas of Mars. It seemed like its pictures were from an entirely different place. It turned out that Mars wasn't exactly a flat, barren wasteland. There were mountains that reached through the clouds, canyons ten miles deep, and huge channels that once carried rushing water.

Oh, What a Beautiful Morning!

On July 30, 1976, Viking I became the first spacecraft to land on Mars. The Viking spacecraft had two parts: an orbiter it used to travel around Mars and a rocket lander to lower itself gently down to the planet's surface. Once on Mars, Viking I measured wind speeds and temperatures. It photographed the sun shining on Mars' surface and collected rock samples. At last, a man-made machine had landed on Mars. Later that year, Viking II put another lander down in an area called the Utopian Plains near Mars' equator. These were the first big steps toward eventually sending humans to the Red Planet.

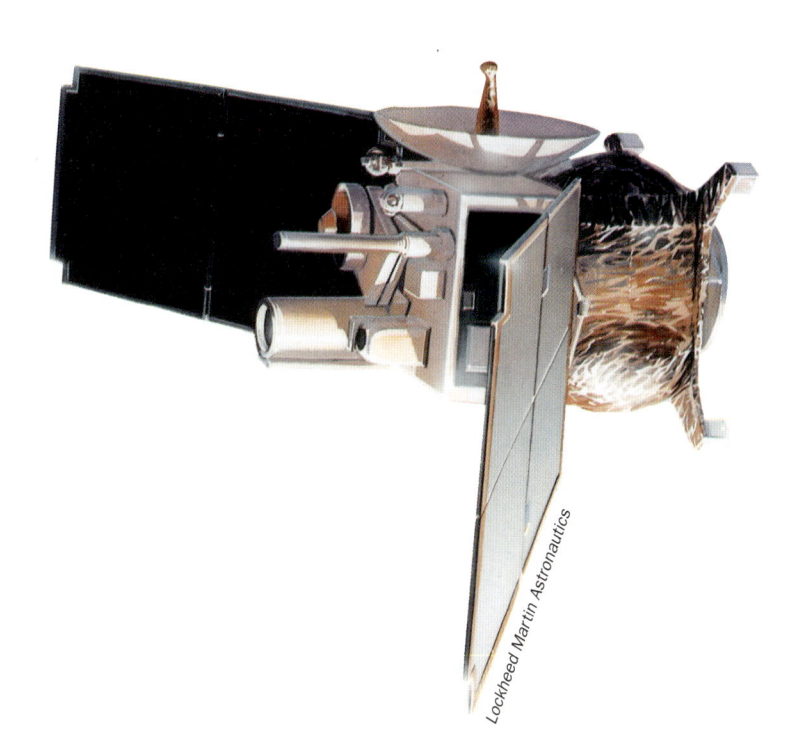

Mars or Bust in 1997!

In the summer of 1997, the Mars Global Surveyor will orbit the planet for more pictures. Then, also in 1997, the Mars Pathfinder satellite will send an automated rover (sort of like a remote-controlled, miniature dune buggy) to Mars. This machine will land at Ares Vallis (Ram Valley), near what some scientists believe is an ancient river bed along the equator. It will take extensive close-up photos of the planet's surface. At chosen sites, the surface will be scraped for chemical tests. The mission of the Mars Pathfinder will be to find out whether water existed on Mars millions of years ago, and, if it did, where.

◀ **This is an artist's drawing of the Mars Global Surveyor. This satellite, which will be launched in 1996, will depend on the world's most advanced technology.**

What Should We Expect on Mars?

Thanks to the Mariner and Viking projects, we know a lot more about Mars than ever before. Exciting secrets are still hidden below the planet's surface, but we know basically what to expect when we land there. Let's look at the problems confronting the terraformers and para-terraformers in their efforts to make Mars suitable for earthlings.

This photo, taken in 1983 by a *Viking* satellite, shows channels that seem to have been created when large amounts of water were released beneath the planet's surface.

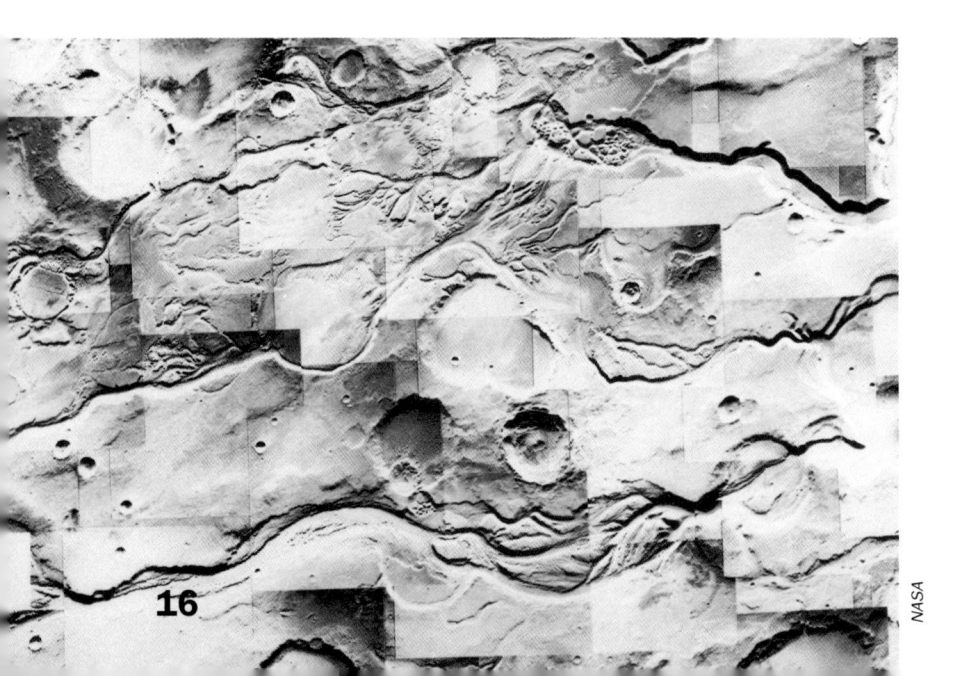

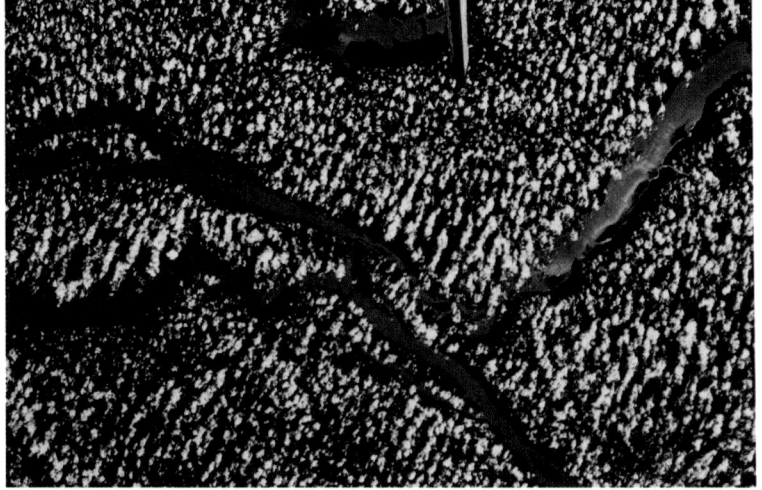

This is a satellite photo of the Amazon river in South America. Much larger and more powerful rivers than the Amazon might have once existed on Mars!

Is There Enough Water?

Without water, no one could live on Mars for very long. Dr. Richard Taylor, a British scientist who studies the different theories of para-terraforming Mars, believes that there is enough water on Mars to support para-terraformers. Taylor says that the existence of glaciers on the planet proves Mars once had a lot of free-flowing water. Even the eighteenth-century astronomer, Giovanni Schiaparelli, had observed river beds etched out in patterns on the planet's surface, indicating heavy water flow. Scientists guess that giant rivers flowed through the Martian landscape about 4 billion years ago. They say that some of these

Martian rivers flowed 100 times faster than the Amazon River! Colonists on Mars would have to figure out how to make Mars' water flow again.

Is There Enough Oxygen?

But even if Mars has enough water to support human colonists, what can we do about the poor quality of Mars' air?

Mars contains only a fraction of the amount of oxygen humans need to live. You would need an oxygen mask if you wanted to breathe there. If we improve Mars' atmosphere, we could make Mars a more breathable place for people.

This is one of the first satellite pictures from the surface of Mars. Do you see any Martians? Does Mars' dry, rocky surface remind you of any places you've visited on Earth?

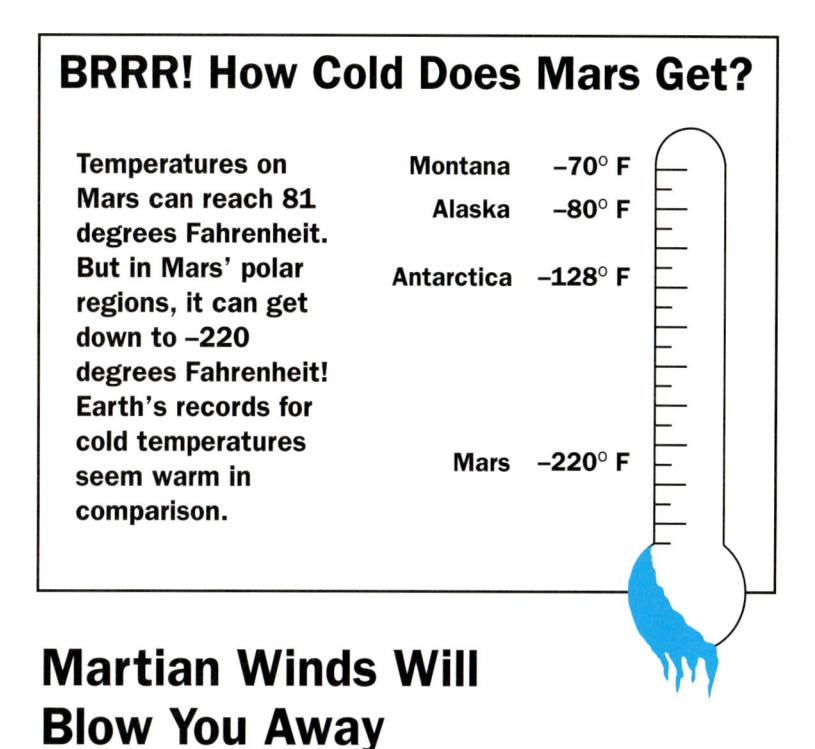

Lockheed Martin Astronautics

BRRR! How Cold Does Mars Get?

Temperatures on Mars can reach 81 degrees Fahrenheit. But in Mars' polar regions, it can get down to –220 degrees Fahrenheit! Earth's records for cold temperatures seem warm in comparison.

Montana	–70° F
Alaska	–80° F
Antarctica	–128° F
Mars	–220° F

Martian Winds Will Blow You Away

Martian winds present another challenge to our attempts to transform Mars. The winds on Mars blow at about 16 miles per hour most of the time. At other times, they can whip up to 73 miles per hour. Unfortunately, there is so much loose, red soil on Mars that when the wind blows, it creates dust storms that can take months to settle down again. Huge, planet-wide dust storms occur in the spring and summer, when Mars is closest to the sun. How could colonists from Earth withstand such terrible storms?

Terraforming: Making Mars a Second Earth

Green plants are essential for providing most of the oxygen we breathe on Earth. What would it take to make mushy, moss-like plants grow on Mars? By improving the atmosphere that probably existed on Mars 4 billion years ago, terraformers believe we could grow these oxygen-producing plants. Once plants begin to grow, the atmosphere would continue to improve.

Other scientists believe the most effective tools for creating this new atmosphere are artificial chlorofluorocarbons (CFCs)—the same substances blamed for making a hole in Earth's ozone layer. Dr. Christopher McKay, a scientist with the National Aeronautics and Space Administration (NASA), wants to fill the Martian air with these CFCs, hoping to cause a "runaway greenhouse effect" on Mars. This would mean the creation of a thick atmosphere around Mars, capable of keeping in heat like a greenhouse.

How would we get these gases to Mars? Eventually, McKay thinks that chemical factories could be built on Mars which could pump gases into the Martian atmosphere.

Mirror, Mirror in the Sky

NASA's Dr. McKay has also suggested the use of mirrors to deliver extra sunlight to Mars, which could speed up the warming process. He claims that mirrors made of large sheets of foil could be thrust into space above Mars. A floating space mirror probably sounds crazy to you, but scientists already have the technology to build it. Such a mirror would be about the size of a cruise ship.

After the sun has reflected off the space mirrors for a few decades, Mars would get warmer. Once this happened, temperatures on most of the planet would become more bearable.

As giant mirrors in space focus sunlight to melt Mars' polar ice cap, huge factories on the planet release special gases into the air, slowly warming the atmosphere. One of Mars' two moons, Phobos, floats in the foreground. ➡

19

Bombs Away!

British scientist Dr. Martyn Fogg thinks that terraformers could use nuclear bombs to cause tremendous underground explosions on Mars. These explosions, says Dr. Fogg, would force water out of the ground, slowly warming up the planet. These bombs would have to be different from any we have right now. In fact, Fogg's bombs would be one million times more powerful than any bomb ever produced on Earth! Dr. Fogg thinks terraformers could use new technologies to reduce the dangerous side effects of the bombs.

Some scientists think that by exploding hydrogen bombs below Mars' surface—like this test bomb exploded in the Pacific in 1952—we could release water hidden in the planet's crust.

The cost of this effort has never been estimated, but it would be enormous. Also, the task of building thousands of super-bombs would be extremely difficult and controversial, even if they were made on Mars.

When Will Mars Look like Home?

Between CFC gases, huge space mirrors, and bombs, how long do terraformers think their job would take?

Most terraforming scientists say it would take thousands of years to warm up Mars enough to make it suitable for simple plants. Once this happened, natural life processes would eventually make Mars more Earthlike. Still, it would be more than 10,000 years before Mars became warm enough or breathable enough for human beings.

Are you willing to wait that long? Most people aren't. It's very difficult to convince people to spend billions of dollars on a project that won't be accomplished—if it's even possible—for 10,000 years. This is the main reason why terraforming will probably never be a solution for those who want to colonize Mars. If we want to make living on Mars a reality, we need to get there much sooner. Maybe someday, after colonies have already been established on Mars, terraforming could take place. This way, people already enjoying the benefits of the new frontier could do the terraforming themselves.

This is where para-terraforming comes in. Unlike terraformers, para-terraformers want to send thousands of people to Mars sometime within the next 100 years.

On a terraformed Mars, colonists could grow plants and raise animals. You could even play outside with your friends—no space suit required!

Para-Terraforming: Building Houses, Not Planets

There are many differences between terraforming and para-terraforming. Terraforming means transforming the climate, air, and entire environment of a planet to make it resemble Earth. In fact, creating a perfectly terraformed planet is like making another Earth, only with different geography.

Para-terraforming, on the other hand, means building structures on the planet where humans could live comfortably, and letting the planet be itself. On Mars, this means setting up buildings people could live in right away without trying to change the planet from being the cold, rocky place it is.

Para-terraforming might mean something as simple as landing a large spaceship on the surface of the planet. After people lived in the ship for awhile, para-terraformers could add more structures, building a larger community little by little. Eventually, an entire city could be built using this method.

This type of para-terraforming is at work in a project called Mars Habitation 2057. Engineered

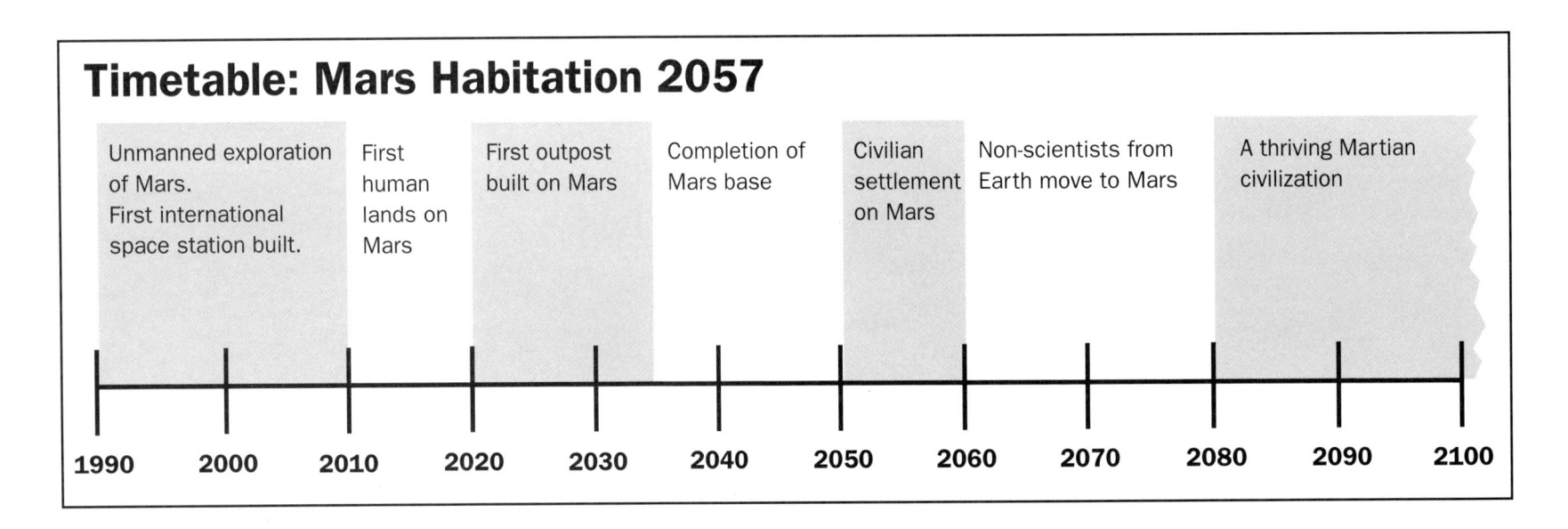

Timetable: Mars Habitation 2057

1990	Unmanned exploration of Mars. First international space station built.
2010	First human lands on Mars
2020	First outpost built on Mars
2040	Completion of Mars base
2050	Civilian settlement on Mars
2070	Non-scientists from Earth move to Mars
2090	A thriving Martian civilization

Habitation 2057 is one example of para-terraforming. Colonists living in this colony could only leave their shelters wearing space suits.

by Japanese scientists Dr. Yoji Ishikawa, Takaya Ohkita, and Yoji Amemiya, the Habitation 2057 settlement would consist of hundreds of inflatable small houses, domes for recreation, and greenhouses for crops. Scientists hope that explorers will be living on Mars by the year 2057.

Getting There

The Mars Habitation scientists think that it will take eight months to travel to Mars. The round trip, which will include some time to explore Mars, will take about two years. That's a long time to stay in one space ship! Russian cosmonauts, though, have spent more than a year in their orbiting space station, *Mir*, so we know extended flights are possible.

The first astronauts traveling to Mars will probably fly in a ship that is longer than a football field, with an interior like the inside of a jumbo jet.

We probably won't take space shuttles to Mars. We will have to build new types of spaceships.

Starting out on the Red Planet

Dr. Ishikawa thinks Habitation 2057 will begin when six to 12 experienced astronauts land on a plain near the Martian equator in the year 2020. According to Dr. Ishikawa, these astronauts' first task will be to get a shelter raised. It will most likely be an inflatable structure, which they will set up as quickly as you might set up a tent. For the first ten years, the astronaut scientists will study the planet, learning how to live in their new environment. There will be flights between Earth and Mars every other year. Ishikawa says that by 2050, fifty explorers will be living on the base, using a solar-powered satellite for energy.

Main Street, Mars

When non-scientists start to arrive after 2050, Mars Habitation 2057 will develop into a community with three distinct types of structures. The homes will be aluminum buildings about as big as large apartments. Each will have a bedroom, a living room, a kitchen, and a bathroom. Each will be buried underneath 11 feet of Martian soil, protecting it from the sun's radiation.

The colonists at Habitation 2057 would use greenhouses and terrariums to grow plants and trees.

Mars Habitat 2057 will also have inflatable greenhouses shaped like sausages cut in half. Inside these buildings, settlers could raise vegetables. The third type of buildings will be large structures called terrariums. Also inflatable, terrariums will be giant, dome-shaped buildings about as big as sports stadiums, where residents could stroll around in an environment similar to Earth's. Fish will swim in the ponds, bees will buzz from flower to flower, even animals like goats and chickens will thrive. Settlers will be able to relax in these terrariums, strolling around the gardens and talking to their neighbors.

An Independent Mars

Planners of Habitation 2057 predict that a population of about 50,000 people will be living in a large, healthy Martian community by the year 2100. Inside their homes and terrariums, the new Martians would feel as comfortable as people on Earth, though they would still need oxygen and spacesuits to leave the shelter. If a family wanted to climb into their Martian rover (a vehicle like a tractor) to see the sunrise, they would still need protection from the harsh Martian environment.

It might be hard for Martian settlers ever to forget that they live in a strange and hostile place. That's why other para-terraformers dream, not of small inflatable houses, but of great buildings whose tops would support a roof over Mars.

All homes would be buried underground for protection against strong winds, freezing temperatures, and the sun's harmful rays.

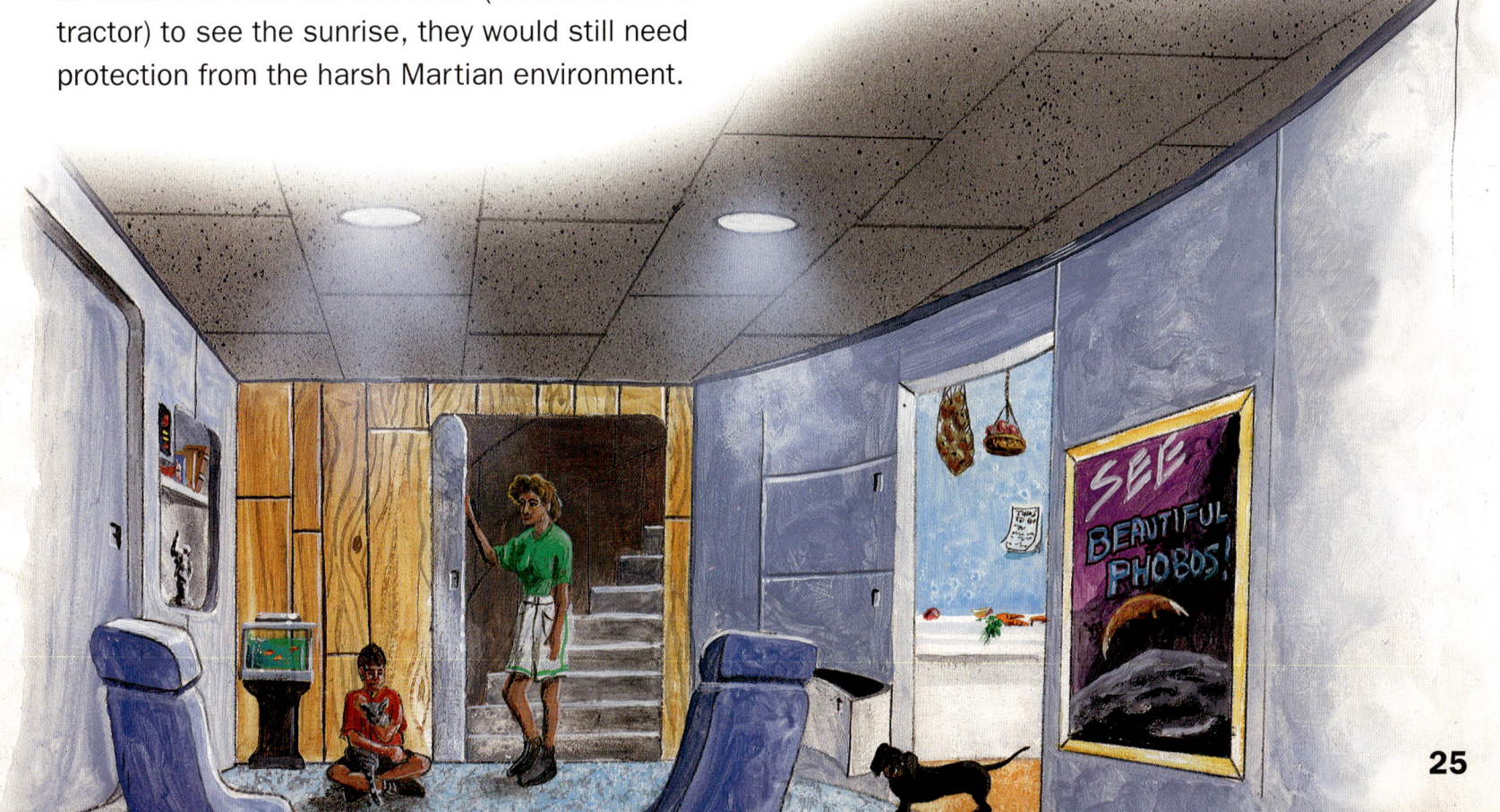

Welcome to Worldhouse!

Inflatable buildings aren't the only examples of para-terraforming technology. Para-terraformer Dr. Richard Taylor, of the British Interplanetary Society (BIS) in England, wants to construct Worldhouse—a two-mile-high, gas-tight roof over most of Mars. Dr. Taylor promises that Worldhouse could eventually provide a home for millions of colonists. He says that people living in Worldhouse wouldn't even realize they were on a strange planet.

Olympus Mons, the highest mountain in the solar system, wouldn't be covered by the Worldhouse roof.

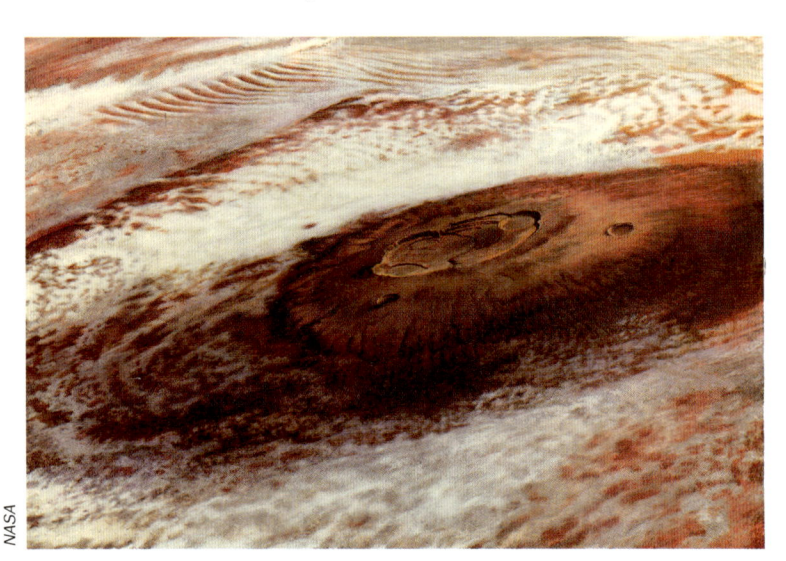

According to Taylor, Worldhouse would be so big that half the area covered by the roof would be in sunlight and the other half in darkness at any one time, just like the north and south hemispheres of Earth. There would even be seasonal changes under the roof. Once the soil had been treated and fertilized, leaves on the trees would turn red and gold in autumn, just as they do on Earth.

Dr. Taylor's plan is to roof 80% of Mars over the next 250 years. According to Dr. Taylor, only Mars' more unusual geographic formations wouldn't be under the roof: the north and south poles, giant mountains like Olympus Mons, and deep canyons like Valles Marineris. Covering most of Mars would create a warm, protected area the size of all Earth's continents put together. That's a lot of land!

Does all this sound too amazing to be true? Maybe. Still, Dr. Taylor says it's possible.

The two-mile-high IMAST (Inhabited Mars Tower Roof Support) would be home to 500,000 people. Eventually, many IMASTs would be built on Mars.

Towers That Touch the Sky

Two kinds of towers would hold up Worldhouse's 2-mile-high roof. The largest and most important tower is the Inhabited Mars Tower Roof Support (IMAST). The IMAST would be 11,480 feet tall—seven times the height of the Sears Tower in Chicago, one of the largest buildings in the world. Is such a building even possible? With modern engineering techniques it is. In fact, Japanese engineers are currently making plans to build the Millennium Tower in Tokyo, which would be about one-fourth the size of IMAST.

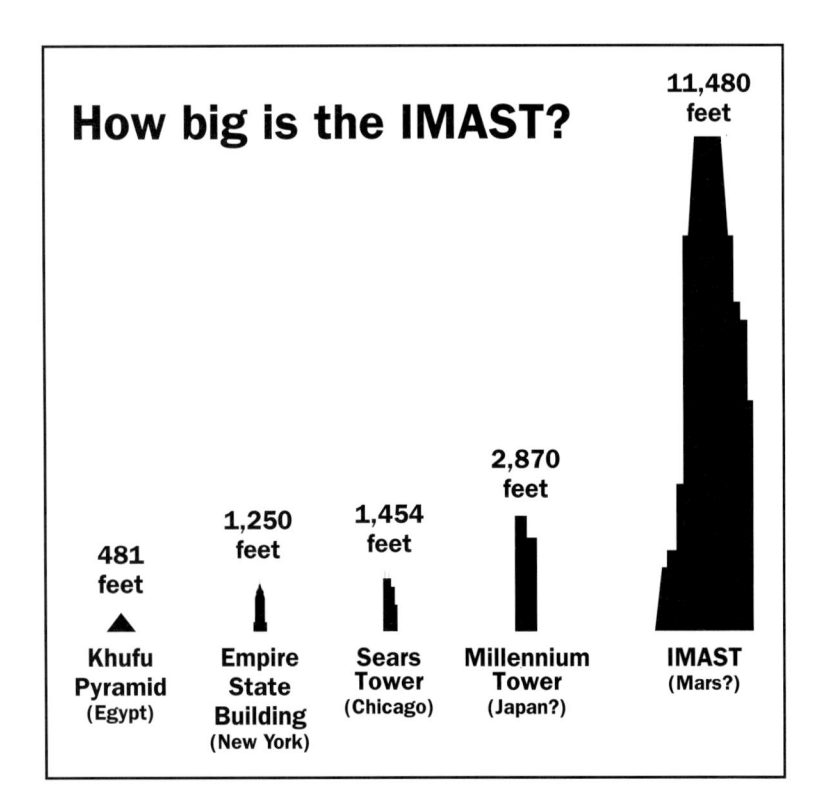

IMAST isn't just a big tower that holds up Worldhouse's roof. It's also a huge skyscraper with a vertical city inside it, big enough for half a million people. Super-fast elevators would carry the residents of IMAST between the building's different levels.

Worldhouse's second type of towers would be the Compression Tension Towers (CTT), more slender towers designed to hold up the roof. Dr. Taylor's plan is to build Worldhouse in sections, with each section requiring one IMAST and 36 CTTs. Each section would be two miles high and cover an area of three miles. The air-tight roof would be made of clear plastic material, so that settlers would still be able to see the pink Martian sky.

Life inside the Worldhouse would be almost like living on Earth. You could bike down the streets, jog, play basketball, or walk the dog. As on Earth, the air inside the Worldhouse would be warm and breathable. Unless the sun shined in a certain way, you probably couldn't even see the plastic roof above you.

How big is the IMAST?

Building	Height
Khufu Pyramid (Egypt)	481 feet
Empire State Building (New York)	1,250 feet
Sears Tower (Chicago)	1,454 feet
Millennium Tower (Japan?)	2,870 feet
IMAST (Mars?)	11,480 feet

Unroofed parts of Mars would be used for mining and solar power plants. In these unroofed areas, residents would still need space suits.

Working Together

Worldhouse would be one of the most expensive, most difficult projects humankind has ever attempted. Building it would have to be a team effort—one country or one nation couldn't do it by itself. The first IMAST would be expensive, but not much more than a freeway from Los Angeles to San Francisco—about one billion dollars. Remember, an IMAST would be able to fit 500,000 people in it. Everything needed to support life would be inside the IMASTs.

One idea of how to protect the Worldhouse's plastic roof would be to intercept falling objects, like meteorites, with guided missiles.

Worldhouse Worries

You might think, "A plastic roof—that sounds flimsy! What if a meteor crashed through?"

This is a concern of Worldhouse scientists, too. However, for many years engineers have been discussing ways to protect Earth from incoming meteors or comets. One idea would be to try intercepting the flying object with a rocket in order to destroy it or change its course. Something similar could be used on Mars.

The Critics Speak Out

Years before there were space shuttles, lunar landers, and communications satellites, members of the British Interplanetary Society (BIS) had already written about or drawn such things. In fact, the BIS motto is "From imagination to reality." Since Worldhouse is probably the BIS' most outrageous dream, it's no wonder the idea is also attracting so many outraged critics.

Critics of Dr. Taylor think that Worldhouse, as well as other theories of terraforming and para-terraforming, are nothing but silly dreams. American scientist, Christopher McKay, has written several articles on terraforming. Even still, he says that we don't know enough about Mars to undertake any major projects in the near future. McKay says that we must first look carefully at the costs and economic benefits of creating human settlements on a desert planet so far away from Earth. When we have all the information, then we can make our decision.

Other scientists say that since we don't even understand our own environment very well, how can we hope to understand the environment on Mars? These scientists say we are foolish optimists if we think we can change Mars quickly.

Dr. Martyn Fogg, who is one of Richard Taylor's former students, is also critical of Worldhouse. "That roof would cost a fortune to maintain," Dr. Fogg says, "and any explosion would leave such a hole, you couldn't seal it before the whole atmosphere was destroyed."

Where Do We Go from Here?

Dr. Taylor admits that the total cost of Worldhouse would be far more expensive and difficult than anything humans have ever attempted. In spite of the staggering price tag and difficulty, Taylor thinks we must eventually live on Mars, if only to make sure the human race survives. He reminds us that a giant meteor probably hit Earth a long time ago, destroying everything in its path, including (perhaps) the dinosaurs. If another monster meteor hit our planet, it could destroy the human race. It would be wise to have an outpost on Mars where we could continue to thrive if something terrible happened on Earth.

The decision about whether or not to move to Mars is up to people like you. Scientists, engineers, and dreamers can only point the way to the Red Planet. It is kids like you who will make the decisions about terraforming, para-terraforming, Worldhouse, and the other fantastic-sounding plans about moving to Mars. Someday, you might even get the chance to make a Mars trip yourself. Who knows? You might become the very first Martian.

What would I say to a dreamer
About to blast off to the stars?
Rev up your main rockets
And drop me off on Mars!

GLOSSARIZED INDEX

This glossarized index will help you find specific topics discussed. It will also help you understand the meaning of some of the words and subjects in this book.

aliens, Hollywood movie, 4
Ares Vallis (Ram Valley)—a valley on Mars near what scientists believe is an ancient river bed along the equator, 15
Aristotle—ancient Greek thinker, 10
astronaut, 23
astronomer, 8
atmosphere, 6, 13, 17, 18
automated rovers—remote-controlled, miniature dune buggies that will explore the surface of Mars, 15
British Interplanetary Society (BIS), 30
canal—a river made by people to move water. Some thinkers thought that the dark lines on Mars were canals built by Martians, 10–11, 12
chlorofluorocarbons (CFCs)—artificial chemicals scientists hope will create an atmosphere on Mars capable of supporting human life, 7, 18, 19, 20
colonists, on Mars, 17, 21, 25, 26
Compression Tension Tower (CTT)—tower that holds up Worldhouse's roof, 28
desert planet, 5, 30
dust storms, 17
Fogg, Dr. Martyn—British scientist who thinks terraformers could use hydrogen bombs to release water from the ground, 20, 30
frontier, 13, 21
Galileo, Galilei—invented the telescope in 1609, allowing astronomers to study Mars more closely, 10
gravity, 5
greenhouse effect—a term used to describe the ability of a planet's atmosphere to keep in heat and oxygen, 18
Herschel, William—eighteenth-century English astronomer who was convinced that Mars was similar enough to Earth to support life, 10
Inhabited Mars Tower Roof Support (IMAST)—an 11,480-foot tower capable of holding 500,000 people, 28, 29
Lowell, Percival—American astronomer, also known as the "canal man," who believed that intelligent beings lived on Mars, 11, 12
Mariner 4—United States satellite that took the first photos of Mars, 14
Mariner 9—United States satellite that flew by Mars in 1971, taking photos that revealed Mars' mountains, canyons, clouds, and channels, 14, 17
Mars ("The Red Planet")—The fourth planet from the sun; a desert planet with no liquid water and no signs of life.
Mars, the ancient Roman god of war, 8
Mars Global Surveyor—a satellite that is scheduled to orbit and photograph Mars in 1997, 15
Mars Habitation 2057—project by Japanese scientists to create a human settlement on Mars by the year 2057, 22–25
Mars Pathfinder—satellite that will send an automated rover to Mars in 1997 to see if water once existed on Mars, 15
Martian—an inhabitant of Mars, 3
McKay, Dr. Christopher—NASA scientist who believes that filling the Martian air with CFCs can create an atmosphere much like Earth's, 18, 20, 30
meteors, 29, 31
Mir—Russian orbiting space station, 23
NASA (National Aeronautics and Space Administration), 18
Olympus Mons (Mount Olympus)—a giant mountain on Mars, 27
orbit, of Mars, 10
ozone layer—layer of gases that protects Earth from most of the sun's dangerous radiation, 13
para-terraforming—building structures on Mars to shelter people against the planet's thin atmosphere and cold temperatures, 22
planets, 3, 6
satellite—a man-made device launched into orbit around a celestial body to gather and send back information to Earth, 14–17
science fiction, 3, 9
Schiaparelli, Giovanni—eighteenth-century Italian astronomer who discovered *canali* (channels) on the surface of Mars, 10, 12, 16
space mirrors—large sheets of foil used to focus sunlight on Mars, 7, 18, 19, 20
solar system, 3
space stations, 12
Taylor, Dr. Richard—British scientist who is one of the leading thinkers involved with the Worldhouse project, 16, 26, 27, 30, 31
telescope, 10
terraforming—transforming the environment of a planet to make it suitable for human life, 6
terrariums—giant buildings on the surface of Mars where plants and animals could live in an environment similar to Earth's, 24
Utopia Planitia (Utopian Plains)—an area of land near Mars' equator, 15
Valles Marineris (Mariner Valley)—a giant canyon in the northern hemisphere of Mars that is more than 2,500 miles long, 27
Viking I and II—the first spacecrafts to land on Mars, in 1976, 14–16
Welles, Orson—actor and director who in 1938 turned the book, *War of the Worlds* into a radio drama broadcast, 9
Wells, H.G.—author of the 1898 book, *War of the Worlds*, which describes a deadly invasion of Earth by angry Martians, 9
Worldhouse—a project that includes building giant towers and gas-tight roofs to provide a home for millions of colonists on Mars, 25, 26